T0194425

essentials

essentials provide up-to-date knowledge in a concentrated form: the essence of what matters as "state of the art" in the current professional discussion or in practice. *essentials* inform quickly, uncomplicatedly and comprehensibly:

- As an introduction to a current topic from your field of expertise
- As an introduction to a subject area that is still unknown to you
- As an insight, in order to be able to speak on the subject

The books in electronic and printed form present the expert knowledge of Springer specialist authors in a compact form. They are particularly suitable for use as e-books on tablet PCs, e-book readers and smartphones. *essentials:* Knowledge modules from economics, social sciences and the humanities, from technology and the natural sciences, as well as from medicine, psychology and the health professions. From renowned authors of all Springer publishing brands.

Goerg H. Michler

Compact Introduction to Electron Microscopy

Techniques, State, Applications, Perspectives

 Springer

Goerg H. Michler
Institute of Physics
Martin-Luther-University Halle - Wittenberg
Halle (Saale), Germany

ISSN 2197-6708   ISSN 2197-6716   (electronic)
essentials
ISBN 978-3-658-37363-4   ISBN 978-3-658-37364-1   (eBook)
https://doi.org/10.1007/978-3-658-37364-1

This Springer imprint is published by the registered company Springer Fachmedien Wiesbaden GmbH, part of Springer Nature.
The registered company address is: Abraham-Lincoln-Str. 46, 65189 Wiesbaden, Germany

What You Can Find in This *essential*

- What distinguishes classical light microscopy from electron microscopy
- What are the techniques and procedures of electron microscopy
- Which requirements must be observed for the sample material
- What methods of sample preparation are used
- How the validity of microscopic examinations can be improved

Preface

For centuries, there has been an interest on the part of mankind in gaining deeper insights into animate and inanimate nature by means of a magnified representation of the respective structures. Initially, this was done by means of simple glass bodies or glass drops held close to the eye and later with improved, compound microscopes. The limited resolution of light-optical devices then led to the development of electron microscopy from the 1930s onwards.

Information about micro- and nanostructures of materials can be obtained in a variety of ways. However, while the scattering methods (small-angle light, small- and wide-angle X-ray, electron or neutron scattering) or the spectroscopic methods only provide averaged (integral) information about extended material areas, electron microscopy allows the direct imaging of structures down to the molecular and atomic level. Pictorial representations offer several contexts simultaneously and stimulate different areas of perception at the same time. The classic saying "A picture is worth a thousand words" also applies to images in science, or as Elias Canetti, the Nobel Prize winner for literature in 1981, put it "The way to reality is through images".

Today, electron microscopy, with a variety of different techniques and methods in addition to the main lines of transmission and scanning electron microscopy, is an indispensable tool in all areas of research, technology and application, both in the material sciences and in the life sciences. For example, processes in the living organism and interactions with drugs can be elucidated in particular with the method of cryo-transmission electron microscopy, the behaviour and properties of implants in the human body can be improved, and the properties of a wide variety of materials can be understood more precisely on the basis of the microstructure by means of analytical electron microscopy and in-situ microscopy, improved in a targeted manner and assessed in a more defined manner with regard to service life or failure safety.

Since the 1960s, Halle (Saale) has been a centre for the use of electron microscopy under the direction of Prof. Dr. Dr. h.c. Heinz Bethge, a centre for the use of electron microscopy was established. Building on this, today, the Heinz Bethge Foundation for Applied Electron Microscopy promotes the use of microscopic techniques in various fields of natural science, biology and medicine. A special concern is the support of young scientists and the arousal of interest in microscopy and electron microscopy and thus in scientific education among students in the upper grades. For this purpose, an extracurricular learning site for electron microscopy has been set up in Halle, where students in the upper grades can also work on projects directly on scanning electron microscopes.

This *essential is* also intended to contribute to the popularization and improved understanding of electron microscopy among colleagues who are more distant from microscopy and interested laypersons. It gives concise information on all techniques and methods of electron microscopy as well as the necessary preparation methods. Selected examples of images illustrate the essential methods of microscopy and preparation. The author's many years of experience in the use of the various techniques and methods of electron microscopy, including probe microscopy, and numerous publications have been drawn upon. The following book are worth mentioning:

- Michler, G. H. (1992). *Kunststoff-Mikromechanik: Morphologie, Deformations- und Bruchmechanismen.* Munich, Vienna: Carl Hanser
- Michler, G. H., & Lebek, W. (2004). *Ultramikrotomie in der Materialforschung.* Munich: Carl Hanser
- Michler, G. H. (2008). *Electron microscopy of polymers.* Berlin, Heidelberg: Springer
- Michler, G. H., & Baltá-Calleja, F. J. (2012). *Nano-and micromechanics of polymers: Structure modification and improvement of properties.* Munich: Carl Hanser
- Michler, G. H. (2016). *Atlas of polymer structures: Morphology, deformation and fracture structures.* Munich: Hanser

They contain clear to detailed descriptions of the various techniques of microscopy and their use for determining the morphology of materials, in particular polymers, as well as for recording micro- and nanomechanical mechanisms in conjunction with the relevant preparation techniques. Current developments of the equipment manufacturers are described in a brochure of the above-mentioned Heinz Bethge Foundation for Applied Electron Microscopy, which was published at the end of 2017 under the title "Electron Microscopy in Halle (Saale) – Status, Perspectives, Applications". For more information on the Bethge Foundation, see www.bethge-stiftung.de.

Halle (Saale), Deutschland   Goerg H. Michler

Contents

List of Figures

Brief History of the Development of Microscopy

The discovery that ground glass could be used to obtain magnified images of the living and dead objects around us led to the use of magnifying glasses with 5–10 times magnification and their designation as "microscopes" from around 1600. Small, molten spheres ~1 mm in diameter then allowed microscopes as early as 200–300 times magnification in the seventeenth and eighteenth centuries. Many important discoveries of microscopic structures were made by numerous users, among them Antoni van Leeuwenhoek (1632–1723). But it was improvements in the melting of optical glass by Joseph Fraunhofer (1787–1826) and above all by Otto Schott (1851–1935) in Jena that made it possible to produce significantly improved microscopes in the mechanical workshop of Carl Zeiss (1816–1888) in Jena and with the collaboration of the physicist Ernst Abbé (1840–1905). With the theoretical explanation of microscopic imaging by Abbé's diffraction theory, there was a change from the pure testing of glass lens combinations that had been common until then to manufacture based on calculated arrangements. Abbé's formula for the maximum possible resolution d of a light-optical system

$$d = 0.61 \ \lambda \ / \ \mathrm{n} \cdot \sin\alpha$$

(d as the smallest distance between two object details that can just be recognized separately; λ as the wavelength of light, α as the aperture angle of the lens, $\mathrm{n} \cdot \sin\alpha$ is the so-called numerical aperture) results in a value of about half the wavelength of the light used ($\lambda/2$). Thus, visible light of a wavelength of 400–800 nm cannot be used to resolve details smaller than about 200 nm (0.2 μm). Based on these theoretical principles, Zeiss microscopes were manufactured, and from about 1870 onwards they became the world's top of the range. Light microscopes and various

G. H. Michler, *Compact Introduction to Electron Microscopy*, essentials, https://doi.org/10.1007/978-3-658-37364-1_1

modifications and improvements conquered applications in almost all fields of technology, natural science and medicine.

Microscopy that is not subject to Abbé's limitation and thus allows better resolutions than in classical light microscopy has only recently been found through work by Stefan Hell on super-resolution fluorescence microscopy, for which he, together with E. Betzig and W. Moerner, was awarded the Nobel Prize in Chemistry in 2014.

The unbeatable advantage offered by direct images of the structures of interest led to the idea of using electron beams in the 1930s as a result of the dual nature of wave and corpuscular beams. Electron beams can be assigned a wavelength λ_e, which is determined by the velocity v of the electrons and hence by the accelerating voltage used ($\lambda_e = h/m_e v$, de Broglie 1924, Nobel Prize 1929) and which at 100 kV is about a factor of 10^5 lower than that of visible light. The realization of this idea was achieved from 1931 onwards by two groups of researchers working independently of each other in Berlin: M. Knoll and E. Ruska at the TH Berlin and E. Brüche and H. Johannson at the AEG. These two groups represented a certain "competitive situation" in which Ruska's group (TH Berlin) pursued the electromagnetic development of the electron microscope and Brüche's group (AEG) pursued the "electrostatic" direction (i.e., the deflection and focusing of the electron beams is achieved by means of current-carrying electromagnetic coils and electrostatic fields, respectively). The first commercially mass-produced **transmission electron microscope** (then called an Over microscope) was introduced in 1939 by Ruska at Siemens with an imaging scale of 30,000:1 and a record resolution of 7 nm (see Fig. 1.1). Until 1945, about 30 devices of this type were manufactured, one of which is in the Deutsches Museum in Munich [1].

In parallel, Manfred von Ardenne (1907–1997) was also working in Berlin in his private laboratory on a magnetic electron microscope (so-called "universal electron microscope"), which he made known to the public in 1940 and which already achieved a resolution of 3 nm [1]. The samples, which were still relatively thick at that time due to the lack of suitable preparation methods, led to a considerable so-called chromatic error,[1] which limited the resolution in addition to other

---

[1] Chromatic aberrations or color errors are caused by the electrons entering an electron lens having slightly different velocities and being focused in different planes near the axis. Thick objects exacerbate this effect due to greater energy losses from the electrons.

Fig. 1.1 View of the "electron overmicroscope" developed by Ernst Ruska, Siemens, about 1939. (From [1])

lens errors. To circumvent this error, M. von Ardenne and M. Knoll worked on a new principle – the scanning principle (analogous to a television tube), which led to a **scanning electron microscope** (then called a "scanning electron microscope") in 1937 [2–5]. Both facilities were destroyed during air raids on Berlin in 1944, ending v. Ardenne's work on electron microscopy.

A rapid further development of electron microscopy took place after the Second World War. In Germany, the Siemens company initially led the way with the ELMISKOP 1 (1954) and the ELMISKOP 101 (1968). From VEB Carl Zeiss Jena, the ELMI D, the best electrostatic microscope ever made (Fig. 1.2), was sold in about 100 copies between 1954 and 1961, and a successful electron-optical system EF was produced in the 1960s.

For a time, electromagnetic electron microscopes were also manufactured at the VEB Factory for Television Electronics in Berlin. In Japan, developments began in 1942 at Hitachi and in 1947 at JEOL, in the Netherlands in 1946 at the Philips

Fig. 1.2 Electrostatic
electron microscope ELMI D,
VEB Carl Zeiss Jena, 1955.
(Source: ZEISS Archive)

company, and in Czechoslovakia mainly after the 1960s at Tesla. Further development of the scanning electron microscope was initially slow until the early 1960s when the first commercial scanning electron microscopes were manufactured in England by Cambridge Scientific Instruments under the name "Stereoscan" with a resolution of 20–50 nm [1] and shortly thereafter by other manufacturers.

The idea of having an electron-optical examination technique for imaging the surfaces of solid objects that cannot be transmitted by light, as in light microscopy, in addition to transmitted-light microscopy with reflected-light microscopy, led to the development of **electron mirror microscopy** in a group led by A. Recknagel [6]. In this technique, an electron beam is directed as a parallel beam perpendicular to the object surface to be examined. By making the potential of the object slightly more negative than the potential of the cathode of the electron beam generator, the electron beam is forced to invert immediately in front of the object surface. The low electron velocity in the reversal region causes a high sensitivity in the detection of field inhomogeneities due to the surface micro relief, but also causes a limited lateral resolving power.

Already in the early days, the vision of a different type of reflected-light electron microscope was pursued, in which electrons emitted from the sample surface lead directly to the desired surface imaging with the aid of accelerator and lens systems. Electron emission can be induced by heating, irradiation with electrons, ions or ultraviolet light. The most widely used variant of emission microscopy today, **photoemission electron microscopy**, has been developed at various locations since the 1950s. In Halle in the 1980s, the group of H. Bethge succeeded in setting up the world's first functioning device that operated under ultra-high vacuum conditions (10^{-10} mbar) and thus enabled defined surface investigations [7].

Another type of microscope is the **field emission microscope** developed by E. W. Müller in the 1950s. The principle is based on the field emission of electrons or ions under a high voltage from an extremely sharp tip and their visualization directly on a luminescent screen without the interposition of lenses. The record resolution of a **"field ion microscope"** at that time was 0.23 nm (imaging scale 1,000,000:1) and even then allowed direct visualization of individual atoms [8]. It clearly exceeded the resolving power of the **"field electron microscope"** and was also more than ten times higher than that of the best transmission electron microscopes of the time.

Ernst Ruska was awarded the Nobel Prize in Physics in 1986 together with G. Binnig and H. Rohrer for their achievements in the development of **scanning tunneling microscopy**. With Ruska's honor, after five decades, the development of (conventional) electron microscopy finally received the recognition it deserved. In the various directions of electron microscopy, scanning tunneling microscopy and more generally **probe microscopy,** although based on completely different physical principles, is often included as a method for direct imaging of submicroscopic structures.

For advances in microscopy, another Nobel Prize in Chemistry was awarded in 2017 to J. Dubochet, F. Frank and R. Hendersen for the development of **"cryo-electron microscopy"**. The combination of cryo-electron microscopes cooled with liquid helium and the process of flash-freezing aqueous substances without ice crystal formation (vitrification) as well as the processing of tilt series, which was hereby honoured, proved to be a breakthrough in the imaging of biomolecules with atomic resolution.

Directions of Electron Microscopy

2

2.1 Overview

Since the inventions in the 1930s, electron microscopy has developed over a wide range with different main directions. These directions can be classified very generally, as in Fig. 2.1, by whether an image is obtained by irradiation as with a "lamp" or by scanning the surface as with a "finger" or a "needle".

The main variants of the electron microscopic imaging technique are:

Type 1: Transmission In **transmission electron microscopy (TEM),** an electron beam emitted from an electron source penetrates the specimen (analogous to conventional light microscopy, but in a high vacuum), and magnified object imaging is achieved with electron-optical lenses connected in series. Ultra-thin samples (in the range of 0.1 μm) are required for this.

Type 2: Reflection or Emission Either a stationary electron beam is reflected from the sample (in high vacuum) **(electron mirror microscopy)** or the sample itself is excited to electron emission by irradiation with electrons, ions or ultraviolet light (indicated by $h\nu$) **(emission electron microscopy, PEEM).** Both techniques can be used to study compact samples.

© The Author(s), under exclusive license to Springer Fachmedien
Wiesbaden GmbH, part of Springer Nature 2023
G. H. Michler, *Compact Introduction to Electron Microscopy*, essentials,
https://doi.org/10.1007/978-3-658-37364-1_2

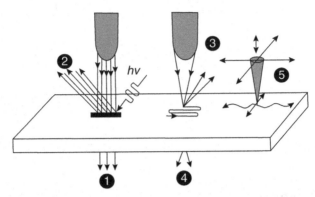

Fig. 2.1 Schematic representation of the main types of electron microscopy. (Adapted from [9], see text)

Type 3: Scanning Beam A focused electron beam is scanned across the sample, producing secondary and backscattered electrons (**scanning electron microscopy, SEM**). In **ambient SEM** or **environmental scanning electron microscopy (ESEM),** the sample chamber is under reduced vacuum or in an environment natural to the sample. Surfaces of compact samples can be investigated advantageously.

Type 4: Scanning Beam in Transmission A focused scanning beam penetrates a thin sample as a variant of type 3, in which the detector is located below the sample that can be transmitted (**scanning transmission electron microscopy, RTEM**). This variant is particularly suitable for use with various analytical techniques and requires thin samples.

Type 5: Scanning Tip A very thin mechanical tip is scanned over the sample and interacts with it based on various physical properties (for conductive samples in **scanning tunneling microscopy** or for insulating materials in **scanning force microscopy**), general **scanning probe microscopy**. A vacuum is not required and surfaces of compact samples can be examined.

2.2 Transmission Electron Microscopy (TEM)

As in (classical) light microscopy, the resolving power depends on the wavelength of the radiation used. In accordance with the dual nature of wave and corpuscular beams, this is determined by the velocity v of the electrons and thus by the accelerating voltage used (de Broglie 1924, $\lambda = h/m_e v$, h = Planck's constant, m_e rest mass of the electrons) (see Table 2.1). The velocity v increases with increasing accelerating voltage U according to the relation $m/2v^2 = e \cdot U$ and reaches at 100 kV 164,000 km/s and at 1000 kV with 283,000 km/s almost the speed of light.

The wavelength of electron beams is about a factor of 10^5 lower than that of visible light (0.4–0.8 μm), but in the transmission electron microscope, due to lens aberrations (especially spherical and chromatic aberration), it allows only about a factor of 10^3 to 10^4 higher resolution. Accordingly, the achievable resolution is in the size range of 0.1 nm and, for maximum resolution devices today, a maximum of 0.05 nm. The development of the resolution in light and transmission electron microscopy over time is shown in Fig. 2.2.

In principle, transmission electron microscopes are constructed in the same way as light microscopes and consist of magnifying lenses connected in series (Fig. 2.3). An insight into the internal structure is provided by a microscope cut open along the central axis in Fig. 2.4.

The electrons emerging from the cathode are accelerated towards the anode and also bundled ("cross over") by the Wehnelt cylinder at the location of the anode. The divergent electron beam is focused by the subsequent electromagnetic lenses (multi-stage condenser lens system) in such a way that it passes through the sample as an almost parallel beam with very little expansion. Since electrons can only propagate in a vacuum, a high vacuum prevails in the column. While in the light

Table 2.1 Relationship between accelerating voltage of the electrons and the resulting wavelength λ_e

Acceleration voltage in kV	20	40	100	200	1000
Wavelength λ_e in nm	0.0086	0.0060	0.0037	0.0025	0.00087

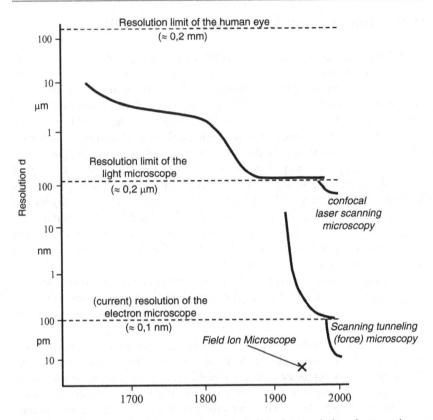

Fig. 2.2 Development of resolution over time in light and transmission electron micros-copy: At the end of the nineteenth century, classical light microscopy had reached its limit until the rapid increase in resolution in electron microscopy began around 1940. (Adapted from [9])

microscope the image can be viewed directly with the eye through the eyepiece, in the electron microscope the electron beams must be made visible on a fluorescent screen. The prerequisite is thin samples that can be irradiated, whose maximum thickness d (in nm) as a guide value for light materials (organic substances and polymers) should not exceed twice the value of the applied accelerating voltage U (in kV) as a first approximation:

$$D[nm] \leq 2 \cdot U[kV]$$

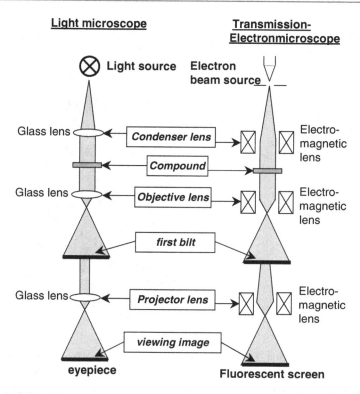

Fig. 2.3 Schematic representation of the beam path in a (transmission) light microscope and a transmission electron microscope. (From [9])

Typical sample thicknesses for such objects are 50–250 nm at 200 kV. The preparation of such objects usually requires a complex preparation technique – which will be discussed in Chap. 3. Thicker samples can be examined in ultra-high voltage electron microscopes (HEM) with accelerating voltages up to more than 1000 kV. However, if high resolution is to be achieved with increasing specimen thickness, the gain in thickness is lower (Fig. 2.5).

The contrast between structural details in amorphous materials is a so-called *"scattering-absorption contrast"* or *"mass-thickness contrast"*. As electrons pass through the sample, electrons are absorbed and scattered (Fig. 2.6). This means that some of the incident electrons are absorbed and deflected in directions different

Fig. 2.4 Former 100 kV
high-resolution microscope
(JEOL 100C), cut open along
the central axis, showing the
internal structure of the
individual pole pieces with
the electromagnetic coils.
(Set up in the student
laboratory "Electron
Microscopy" of the Heinz-
Bethge-Foundation in Halle
[Saale], own photo)

from the beam direction. The scattered electrons are collected by a contrast aperture
(aperture diaphragm, at the lower focal point of the objective lens) The fraction of
electrons that are scattered to larger angles depends on the atomic number of
the scattering atom (i.e. the density of the material) and the local thickness of

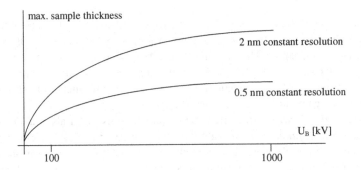

Fig. 2.5 Increase in maximum sample thickness with accelerating voltage at constant resolution. (Schematic, from [9])

Fig. 2.6 Schematic representation of the mass-thickness contrast: thin (d₁) and light materials (density ρ₁) scatter electrons less than thick (d₂) and heavier object sites (density ρ₂) and appear brighter in the image. (From [10])

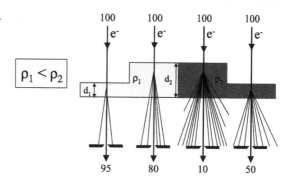

the sample. The higher the atomic number and the sample thickness, the greater the fraction of electrons scattered to larger angles. These electrons are missing from the image formation on the fluorescent screen and result in darker image areas (Fig. 2.7).

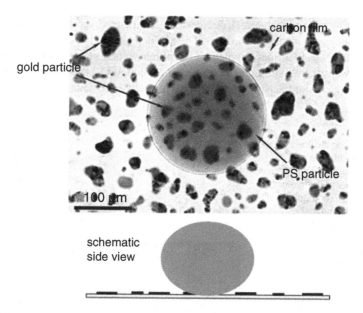

Fig. 2.7 Schematic representation of the mass-thickness contrast on a model object: Thicker polystyrene particles and smaller but heavier gold particles scatter electrons more strongly than the carbon carrier film and appear darker in the image. (Adapted from [10])

By means of the rays transmitted through the contrast diaphragm, the objective lens produces a one-level magnified image of the specimen. This first intermediate image is strongly magnified by the subsequent intermediate lenses and finally by the projector on the fluorescent screen (the intermediate lenses are not shown in the diagram in Fig. 2.3). The fluorescent screen converts the invisible electrons into visible light, and the image is recorded by photographic plates or a camera.

For specimens with hardly any differences in density and object thickness, the contrasts can be significantly enhanced by selective staining techniques (e.g. incorporation of heavier atoms such as osmium or ruthenium in one phase of the specimen) (see Sect. 3.3).

For crystalline samples, *diffraction contrast* is also used, whereby bright-field and dark-field images are possible through the use of the contrast diaphragm (see Fig. 2.8). In a similar way to X-rays, electron beams are diffracted by crystalline samples according to Bragg's law, leading to dot diagrams for single crystals and ring diagrams (Debye-Scherr rings) for many-crystal systems. From both types of diagrams, the lattice parameters or the lattice plane distances of the materials can be determined.

One way to improve the resolution is to reduce the wavelength by increasing the accelerating voltage (Table 2.1). A higher accelerating voltage also allows thicker usable samples compared to the limited penetration capability of 100 kV electrons. Both reasons led to the development of **ultrahigh-voltage electron microscopy** at 1–3 MeV (Fig. 2.9). Substantially improved resolution was then soon no longer a focus, as the use of thicker samples brought a variety of advantages in the study of

Fig. 2.8 Crystalline structures in a semi-crystalline polymer (LDPE – low density polyethylene) with a sheaf morphology of bundle-like ordered crystalline lamellae. (**a**) In bright field (crystalline lamellae appear dark), (**b**) in dark field (lamellae appear light), (**c**) in diffraction pattern (corresponding to a single crystal diagram). (From [9])

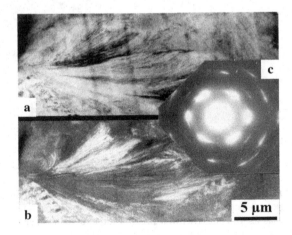

Fig. 2.9 High voltage electron microscope (HEM) JEM-ARM1300S (Atomic Resolution Microscope) with an accelerating voltage of up to 1,300,000 V (1.3 MV). (Source: JEOL [Germany] GmbH, Freising)

metals, ceramics, polymers, and biological and medical materials. In particular, **in-situ microscopy,** in which the influence of external factors such as temperature, mechanical forces or electromagnetic fields on material changes is investigated, opened up new fields of application in materials science (see Sect. 2.6). The realized limit for ultrahigh-voltage electron microscopes (HEM) is 3 MeV, since above this level the gain in usable sample thickness decreases, but the costs increase disproportionately. The first 1 MeV HEM in Germany was installed by JEOL on the initiative of H. Bethge at the institute in Halle (Saale).

From the beginning, there was a constant search for ways to reduce the sources of error in electron microscopic imaging (aberration, instabilities) in order to achieve improved resolution of electron microscopes. In the 1980s, the resolution of the best instruments was 0.2 nm. In **high resolution electron microscopy (HREM)**, aberration correctors are used to correct two types of error: lens spherical error due to variation in refractive power across the lens cross-section (Cs correction) and chromatic error due to variations in accelerating voltage during beam generation or in thick samples (Cc correction). The maximum resolution achievable today in corrected (S)TEM is 0.05 nm, cf. Figure 2.2 [11]. However, this is still far from the maximum resolution due to the wavelength of the electrons. Next steps are coherent electron sources and further higher aberration corrections.

Further developments of transmission electron microscopy are electron holography and energy filtering electron microscopy. When a wave (the electron wave) passes through an object, it is modulated in amplitude and phase. The square of the amplitude is visible in the TEM image as a contrast variation, while the phase remains largely invisible. In **electron holography**, the wave is recorded in an interference pattern (the hologram) and completely represented in terms of amplitude and phase using the steps of numerical image processing. In the end, an amplitude and a phase image are obtained [12]. In particular, the phase image, invisible in the normal TEM, is interesting because it allows access to the internal fields (electric and magnetic nanofields) that account for a significant fraction of structure-property relationships of a solid.

Further information from the electron-object interactions is acquired with **analytical transmission electron microscopy**. For conventional TEM, only the elastic scattering processes of the imaging electrons in the sample are significant (mass-thickness contrast cf. Figure 2.6). Due to inelastic interactions, in which electrons of the electron beam not only undergo a change of direction but also suffer a loss of energy, these electrons are lost to the imaging process because their energies (wavelengths) are no longer matched to the set parameters of the lenses and correctors. Therefore, the inelastically scattered electrons merely create a diffuse scattering background in the image and thus reduce the image contrast and also the resolving power. The additional exploitation of the inelastic interactions for an information gain is done by **energy filtering transmission electron microscopy (EFTEM)**. The separation of electrons without energy loss (zero-loss) up to an energy loss of several 10 eV is achieved by filter systems. In practice, filter systems from the companies Gatan ("Gatan Imaging Filter" – GIF) and LEO ("Omega" filter) have become established (see Fig. 2.10). The advantages lie in local chemical analysis. One advantage of zero-loss imaging is the possibility of examining thicker samples with improved resolution, since the inelastic interactions that occur here between the electron beam and the sample are eliminated by the energy filter (see Fig. 2.11).

Electron irradiation mainly damages organic materials (such as biological, medical materials but also polymers) to a greater or lesser extent. Primary processes occur during irradiation, such as ionization and the breaking of chemical bonds. Secondary processes, such as chain scission, loss of mass (desorption), loss of crystallinity, cross-linking, sample heating, electrical charging and even sample movement, have a stronger influence on the examination. As a rule of thumb, the lower the carbon content, the greater the sensitivity to irradiation, i.e. for polymers the sensitivity increases in the order PS, PE, PC, PMMA, PVC, PTFE. Since the

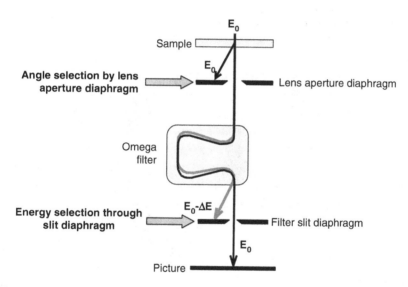

Fig. 2.10 Principle of energy-filtering TEM (EFTEM) with the so-called omega filter (LEO, Zeiss), in which inelastically scattered electrons can be excluded from the imaging process (zero-loss imaging is shown). (From [9])

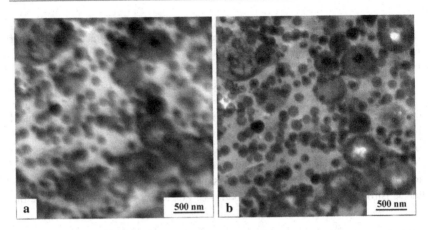

Fig. 2.11 TEM images of a thick polymer sample (ABS: impact modified polymer by rubber particles [dark stained with OsO₄], section thickness 400 nm): (**a**) in conventional imaging and (**b**) with zero-loss imaging in EFTEM. (From [9])

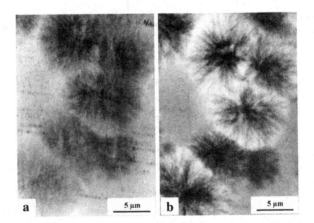

Fig. 2.12 Spherulites with radial fibril structure in a polyurethane, (**a**) little irradiated (**b**) more irradiated with contrast development of the radial structure in the spherulites. (From [10])

damage occurs at the molecular level, this does not necessarily affect the morphology analysis. Sometimes irradiation-induced material damage even contributes to a beneficial contrast enhancement, as by selective crosslinking in semi-crystalline polymers (see Fig. 2.12, cf. also Fig. 3.10) [10]. The rate of damage to samples in the electron microscope can be reduced by a number of experimental means [9, 10]. One instrumental way to reduce irradiation damage is to cool the specimens using special cooling tables. Cooling down to the temperature of liquid nitrogen (−196 °C) is common, and for special investigations, e.g. of bacteria, down to the temperature of liquid helium in the **cryo-transmission electron microscopes.**

2.3 Scanning Electron Microscopy (SEM)

In the scanning electron microscope, the image is not generated by lenses as a whole, but is assembled line by line, as in television technology. The schematic structure of a scanning electron microscope is shown in Fig. 2.13 with the main assemblies.

The electron gun generates an electron beam which is focused as a very small spot on the sample surface in a high vacuum by the subsequent electromagnetic lenses. By means of deflection coils, this electron beam is guided line by line (scanning pattern, analogous to the structure of a monitor image) over a selected sample area. Electrons are emitted from the sample, collected in a detector and amplified. Synchronously, an electron beam is moved in the image tube of the evaluation unit of the SEM. Each pixel visible on the SEM monitor corresponds to a precisely defined point on the sample surface. The magnification can be easily adjusted by the ratio of the screen size and the size of the sample area scanned by the electron beam (e.g. 10 cm screen size to 0.1 mm sample size corresponds to a magnification of 1000:1).

When the primary electron beam strikes the sample, elastic and inelastic scattering processes with the atoms of the sample give rise to several interaction products, the secondary (SE) and backscattered electrons (RE) mainly used for image formation and the X-rays used for chemical analyses (see Fig. 2.14). The size of the area of interaction between the electron beam and the sample (interaction volume) depends on the energy of the primary electrons (accelerating voltage) and the atomic number of the sample material (see Fig. 2.15). The higher the accelerating voltage and the smaller the average atomic number of the sample site, the larger the interaction volume. The formation of contrast by means of secondary electrons (SE) results essentially from the angle of inclination of the respective sample site to the incident primary electron beam: In the case of perpendicular incidence, fewer secondary electrons are emitted from the sample than in the case of oblique irradiation, so that oblique sites and edges appear bright and very contrasty surface structures are imaged *(edge effect, topography contrast)* (see Fig. 2.16). The number of backscattered electrons (RE) depends strongly on the material of the

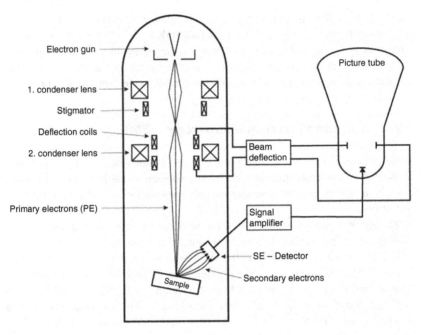

Fig. 2.13 Schematic diagram of the structure of an SEM. (From [9], see text)

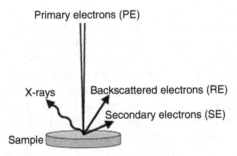

Fig. 2.14 Interaction products of the primary electron beam (PE) with the sample surface. (From [9])

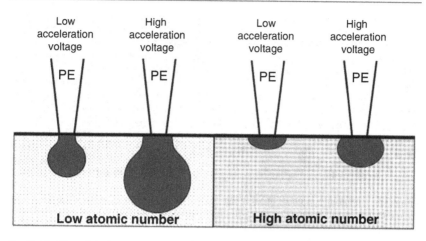

Fig. 2.15 Size of the interaction volume between incident primary electrons and the sample as a function of accelerating voltage and atomic number of the sample material. (From [9])

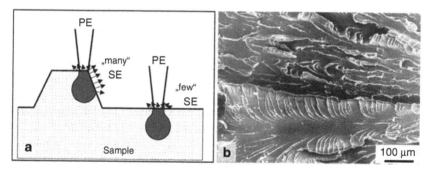

Fig. 2.16 Edge effect in secondary electron imaging in the SEM: (**a**) schematic representation (**b**) high-contrast representation of fracture edges in a polymer (polycarbonate). (From [9])

investigated sample site and increases with the atomic number. Therefore, RE imaging is particularly suitable for investigations of the material composition of a sample *(material contrast)*. Figure 2.17 shows a polymer matrix with inorganic filler particles (bone cement) in the secondary electron and backscattered electron

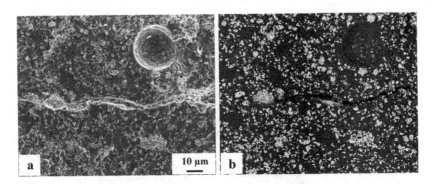

Fig. 2.17 Topography contrast in SE image (**a**) and material contrast in RE image (**b**) of the same site in a bone cement. (Polymer with filler particles, from [9])

image: the SE image mainly reflects differences in topography, while the RE image shows the inorganic particles particularly clearly due to their greater density (atomic number).

As a further interaction of the electron beam with the sample, X-rays are produced (see Fig. 2.14), the energy analysis of which provides further information of the material composition up to quantitative chemical analysis. The X-rays are characteristic of the elements contained in the sample and can be analysed either in terms of energy (energy dispersive analysis of X-rays – **EDX**) or wavelength (wavelength dispersive analysis of X-rays – **WDX**). X-ray analysis can be used to determine not only the presence of elements, but also their spatial distribution and relative abundance. Figure 2.18 shows inorganic particles in a particle-filled silicone rubber (catheter made of silicone with X-ray contrast agent) on the fracture surface and the elements C (originating from the conductive sample vaporization), O and Si (originating from the silicone matrix) and Bi in the EDX spectrum. The comparison of the SE image (Fig. 2.18a) with the Bi distribution from the same sample location is called *element mapping* and shows that the particles on the fracture surface are the particles of the X-ray contrast agent.

SEM combined with EDX is now the standard method in the elucidation of structures and chemical composition. Conductive materials can be investigated without any further pre-treatment, whereas a conductive coating (e.g. vapor depo-

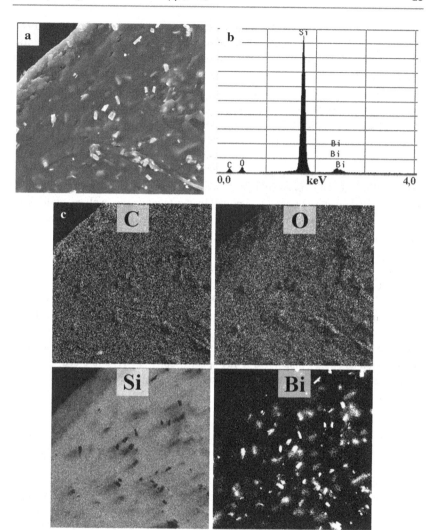

Fig. 2.18 Cross-section through a particle-filled silicone rubber (catheter with X-ray opacifier) in the SE image (**a**), with the EDX spectrum (**b**) and the distributions of C, O, Si and Bi (**c**, element mapping). (From [9])

sition with a thin carbon layer) is required for non-conductive samples. This can be dispensed with in a so-called ambient or low-vacuum SEM (**Environmental SEM – ESEM**), in which it is possible to work with a higher partial pressure (up to a pressure of several Torr) in the sample chamber with the aid of a special aperture system. The gas atoms in the sample chamber are ionized by the electron beam, and the positive ions compensate for the negative charge on the sample surface caused by the electron beam. This is also a great advantage for micromechanical in-situ investigations of electrically non-conductive materials (see Sect. 2.6). By lowering the temperature in the sample chamber to a few degrees Celsius, a humidity of 100% can be achieved, which completely prevents moist samples from drying out. This opens up new areas for electron microscopic examination, such as moist, water-containing samples in the technical as well as biological-medical fields. An example is shown in Fig. 2.19 of a glass ionomer cement for the dental sector, where an examination at higher pressures shows the material as in the application area (Fig. 2.19a), while in the SEM or at low pressures drying out with separation of the constituents (Fig. 2.19b, drying cracks) occurs, which, however, does not correspond to the practical case.

In addition to the secondary and backscattered electrons detected in the SEM, the scanning beam can also penetrate thinner samples and be detected by a detector located below the sample (**type 4** in Fig. 2.1). This **scanning transmission electron microscopy (RTEM)** is often combined with various analytical techniques such as energy dispersive X-ray spectroscopy (EDX) and electron energy loss spectroscopy (EELS) and achieves atomic resolutions via special dark field techniques.

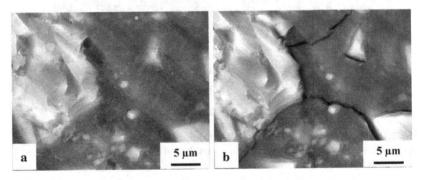

Fig. 2.19 ESEM images of a glass ionomer cement at 4 °C and varied water vapour pressure: (**a**) 5.9 Torr, (**b**) 2.2 Torr with shrinkage (phase separation, shrink cracks). (From [9])

Usually, electron accelerations of 10–30 kV are used in SEM. Here, chemical bonds are broken in radiation-sensitive materials, such as polymers. Reducing the accelerating voltages, i.e. the electron energies, allows the investigation of undamaged structures at high magnifications. This **low-voltage SEM** and **low-voltage high-resolution scanning electron microscopy (LV-SEM)** also provides particularly high-contrast material differentiation in the thinnest surface layers. At 0.3 kV, resolutions of better than 0.7 nm are currently achieved in the secondary electron image [11].

In addition to improvements in spatial resolution and better analytics, recent developments also aim to increase the imaging speed. Acceleration of imaging by faster deflection of the primary beam would require faster detectors to measure the secondary electron signal and a higher current intensity of the primary beam, which degrades the resolution as the electrical repulsion of the electrons in the primary beam widens the focus. A possible solution is to use many electron beams simultaneously in a single setup, in a so-called **multibeam SEM** [13]. After various prototypes, there is a practical device with the MultiSEM from ZEISS [14]. Here, instead of using a single primary beam, the sample is illuminated with a bundle of many primary beams simultaneously. A common projection optics with deflector focuses and scans the beam over the sample. The parallel use of multiple electron beams (over 91 in a newer instrument version) increases the sample area that can be imaged in one raster pass. The high degree of automation allows the acquisition of hundreds of serial sections of a sample in a short time (as can be obtained with an ultramicrotome with automatic section collector – see Sect. 3.1) and opens up many future possibilities (see Fig. 5.1 and Chap. 5).

2.4   Reflection and Emission Electron Microscopy

As already mentioned in Chap. 1, the idea of having an electron-optical examination technique for imaging the surfaces of solid objects that cannot be transmitted by light, as in light microscopy, as a supplement to transmission microscopy with reflected light microscopy, led to the development of electron mirror microscopy and emission microscopy (cf. **type 2** in Fig. 2.1), mostly **photoemission electron**

microscopy (PEEM). Parallel to the PEEM development was the development of a reflection electron microscope based on surface-sensitive diffraction of slow electrons. This technique, known as **low-energy electron microscopy (LEEM)**, is usually operated in combination with PEEM and also allows electron-mirror microscopic imaging of surfaces [15].

The UHV-PEEM technique introduced by Prof. Bethge marks the starting point of many further developments and applications of electron microscopy on solid surfaces. For his work on imaging catalytic surface reactions on platinum using UHV-PEEM, G. Ertl was awarded the Nobel Prize in Chemistry in 2007. Further progress has been made with the application of novel light sources for photoemission; for example, the use of ultrashort laser pulses allows PEEM imaging of dynamic processes on surfaces with femtosecond time resolution [16]. LEEM-PEEM is usually operated under UHV conditions. Surface reactions also under approximate normal pressure conditions (NAP – near ambient pressure) can be studied with a new NAP-LEEM-PEEM [17].

2.5   Scanning Probe Microscopy

In the early 1980s, an original approach by G. Binnig and H. Rorer at IBM in Rüschlikon, Switzerland led to the development of a completely different type of microscopy, **scanning tunneling microscopy** [18]. Here, a fine metal tip is scanned very closely over the sample surface using piezoelectric actuators. A voltage is applied between the tip and the sample so that a tunnel current flows as a result of the very small distance. This is used for line-by-line imaging and allows surface representations down to the atomic level. In addition to the tunnel current (for conductive samples), various other physical interactions between the sample and the tip can be used for non-conductive materials to form images in scanning force microscopes or in **scanning probe microscopes** in general. In Scanning Force Microscopy (**SFM or atomic force microscopy, AFM**), various forces (atomic forces, van der Waals forces) are used as interaction mechanisms. Such a microscope is quite simple in principle, does not take up much space, but requires extensive data processing. Figure 2.20 shows an atomic force microscope in overview (Fig. 2.20a), sectional view (Fig. 2.20b) and as a principle sketch (Fig. 2.20c) [10]. The fine metal tip is deflected a little by interactions with the sample. The deflection is analyzed by a laser beam. The sensor tip is attached to a miniaturized leaf spring (cantilever), which is bent according to the effective force up to a maximum of about 10 nm if the distance between tip and sample is sufficiently small. The deflection of a laser beam reflected from the back of the spring is measured with a four-quadrant detector.

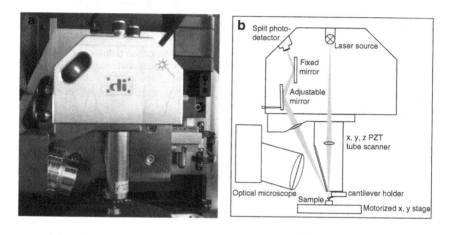

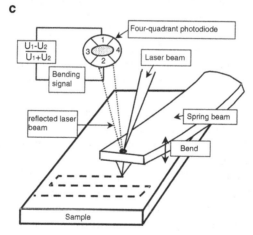

Fig. 2.20 Principle of an atomic force microscope (AFM) for the detection of distance-dependent force interactions between tip and sample, (**a**) view (**b**) sectional view (**c**) principle. (Adapted from [9, 10])

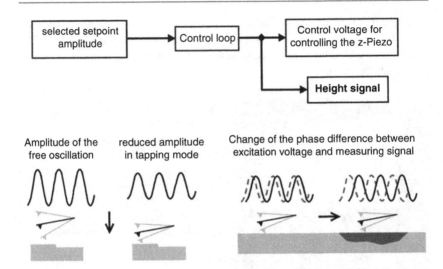

Fig. 2.21 Schematic representation of the "Tapping Mode". (From [9])

Particularly advantageous here are techniques for dynamic force microscopy, in which the spring beam is set in oscillation perpendicular to the sample surface and changes in amplitude, phase or resonant frequency of the spring beam oscillation are evaluated. The "tapping mode" has become particularly important, in which the distance between the sensor tip and the sample surface and the vibration amplitude are set in such a way that the tip only touches or penetrates the sample surface for a fraction of the vibration period during each vibration cycle (see Fig. 2.21). In addition to a topography image, local stiffness changes of the sample surface (phase signal image) can thus be determined.

Figure 2.22 shows, in comparison with a TEM image (Fig. 2.22a), a phase image of a heterogeneously structured polymer (Fig. 2.22b) consisting of a partially crystalline, lamellar HDPE matrix (high density PE) with VLDPE particles (very low density PE). At object locations of lower stiffness (softer material) larger vibration amplitudes are registered, which are reproduced in the image as lighter structures: soft VLDPE particles in the matrix (they are dark in the TEM image due to strong contrast) and amorphous portions between the lamellae (they appear grey

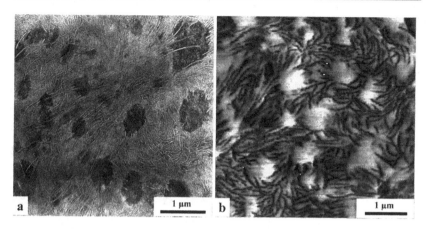

Fig. 2.22   TEM image (**a**) and AFM phase image (**b**) of a heterogeneous polymer blend of a lamellar PE matrix with soft particles. (From [9])

in the TEM image due to lower contrast). Although the general structure rendering is the same in both cases, the AFM image shows a further detail with a line-like arrangement of small crystals (mosaic blocks) within the dark lamellae, which is not visible here in the TEM image.

2.6   In-Situ Microscopy

Electron microscopic techniques can be used to investigate not only quasi-static structures in materials, but also dynamic changes in the material under various environmental influences – this is the field of **in-situ microscopy.**

For the analysis of micro- and nanomechanical processes, there are different strain and bending apparatuses for SEM, TEM and AFM [10] (see Fig. 2.23).

In-situ microscopy is particularly effective in the ultrahigh-voltage electron microscope, since here, due to the high beam voltage, the specimen thicknesses that can be examined are so large that the material properties often already correspond to the properties present in the compact body [19, 20]. Figure 2.24 shows a series

Fig. 2.23 Tensile device
from Oxford Instruments,
which allows in-situ strains in
the SEM in the temperature
range −180 to +200 °C.
(Adapted from [10])

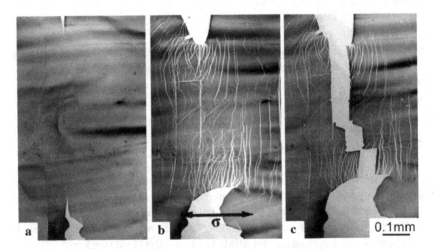

Fig. 2.24 In-situ deformation of a glassy polymer (0.5 μm thick PS thin section) in 1000 kV
HEM: (**a**) The initial sample is notched at the top and bottom; (**b**) under load, so-called
crazes have formed starting from the notches perpendicular to the strain direction σ; (**c**) increasing load leads to crack initiation and tearing along the crazes. (From [21])

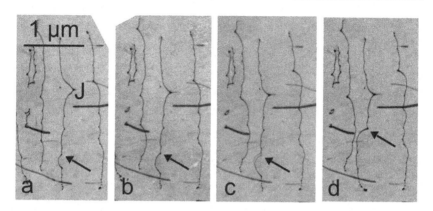

Fig. 2.25 In-situ strain test in 1000 kV HEM: Successive stages of dislocation motion (dark lines) during deformation of a MgO crystal; under load they bulge and move (see arrows). (Courtesy of U. Messerschmidt, F. Appel, 1987)

of successive HEM images of a glassy polymer with crazes[1] (bright bands) formed under stress as precursors to the subsequent cracking of the sample [21]. Substantial advances by in-situ microscopy have also been found on semiconducting materials, metallic and ceramic crystals, in that such tests show the movement of dislocations[2] (see Fig. 2.25). With tailored apparatus for the 1000 kV HEM, samples can also be loaded at higher temperatures (up to over 1000 °C) for defined high temperature strain tests) [22].

In addition, thermal effects can be investigated in-situ with heating holders and cooling tables, and electrical and magnetic effects can be investigated by applying microfields. Vacuum-tight humidity cells ("closed cell" in-situ TEM holders) allow moisture-containing and biological materials to be studied in a natural environment (an alternative newer technique is the ambient SEM or ESEM – see Fig. 2.19).

---

[1] The crazes are local deformation zones of several 100 nm thickness and several μm length, which essentially cause the fracture toughness.

[2] Dislocations are lineal crystal defects and carriers of plastic deformation of crystals.

Sample Preparations

3

3.1 Overview

Today, the various microscopic techniques allow structural elucidation down to the subatomic size range. In order to exploit this possibility, special preliminary steps or preparations to improve the structural visibility or contrast are usually required. The scheme in Fig. 3.1 illustrates which techniques can be used to investigate surfaces and which can be used to analyse the interior of materials. With scanning electron microscopy (SEM) and atomic force microscopy (AFM) and further techniques of scanning probe microscopy, surface structures can be directly analysed. The main method used for the study of internal structures is the preparation of thin and ultrathin samples and their study in TEM and AFM. The preparation of replicas and their examination in the TEM was used more often in the past and has now been replaced by the much simpler SEM and AFM [9, 10].

The optimal use of electron microscopic techniques requires sample preparations that are adapted to the material to be examined. The old saying "well prepared is half microscoped" still applies. Figure 3.2 shows an overview of procedures for the preparation of surfaces with elaborated structural details, preparation of thin layers of mainly harder, inorganic materials as well as thin sections of softer (soft matter) and biological materials.

© The Author(s), under exclusive license to Springer Fachmedien
Wiesbaden GmbH, part of Springer Nature 2023
G. H. Michler, *Compact Introduction to Electron Microscopy*, essentials,
https://doi.org/10.1007/978-3-658-37364-1_3

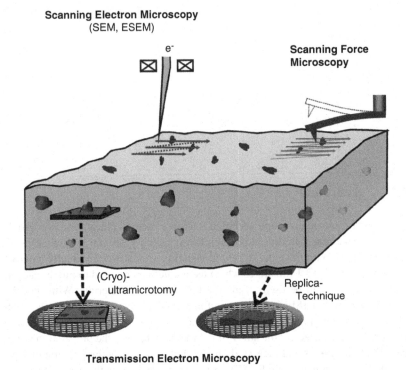

Fig. 3.1 Use of the various microscopic techniques to examine surfaces and the interior of materials. (From [10])

These main lines include, roughly outlined, the following preparation techniques:

- *Preparation of surfaces:* Only rarely does a freely grown surface show the material structure. More informative are smooth (ultra-)microtome sections (see Sect. 3.3) or sections that can be selectively etched (by chemical or physical means) to enhance contrast. A rarely used indirect method is the replication with transmission electron microscopy (TEM). In simple cases, the internal morphology can also be exposed via defined fracture surfaces and recorded in the SEM.

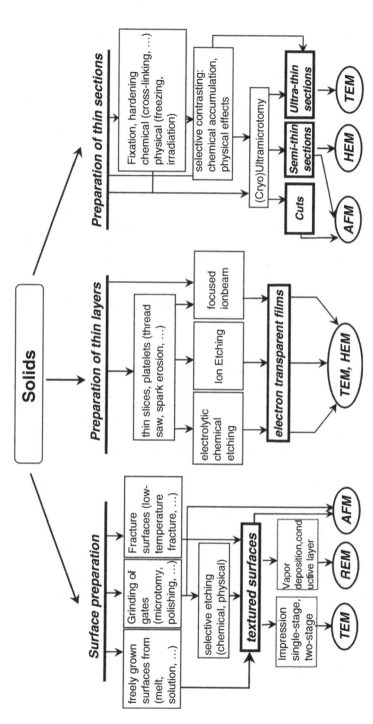

Fig. 3.2 Overview of main lines of preparation and electron microscopic investigation possibilities for the determination of the morphology of compact materials. (From [9])

- *Preparation of thin layers:* Thin slices with thicknesses as close as possible below 0.1 mm, mostly produced by mechanical methods (thread sawing, grinding, polishing) or ion beams, are thinned in a second step by electrolytic, chemical etching or ion etching to such an extent that electron transparent areas are created. Depending on the material density, the required film thicknesses for the conventional 100–200 kV TEM range from a few 10 nm to a few 100 nm and for the 1000 kV HEM up to a few micrometers. For the focused ion beam technique, see Sect. 3.2.
- *Preparation of thin sections:* The method that is universally applicable, especially for softer, organic materials, is (cryo) ultramicrotomy. Cryo- and ultramicrotomy are used to produce semi- and ultrathin sections of compact materials for examination in the TEM, SEM or AFM. Harder materials such as hard polymers and soft metals can be cut directly. Softer materials such as most polymers and bio-medical materials must be hardened ("fixed") or frozen to lower temperatures in the cryo-ultramicrotomes before cutting. In addition to hardening (fixing) the material, contrast enhancements are often required.

3.2 Preparation of Thin Films (Using Etching Techniques)

Etching techniques play the central role in the investigation of inorganic hard materials. Their selection and execution determines the overall effort of the investigation, but also the achievable informative value of the structure determination, especially also in high-resolution electron microscopy. It must be ensured that no mechanically or thermally induced damage artefacts are produced in the sample by the etching process or that the structural details of interest are not obscured. Nowadays, **ion beam etching techniques** play a central role. Ions are extracted from a source, bundled into a beam by electrostatic fields and simultaneously accelerated towards the sample. Due to the mechanical momentum transfer of the impinging ions, a defined material removal is produced without the need for mechanical stress (as is the case with metallographic grinding methods). The most commonly used ion beam technique is the **focused ion beam technique (FIB)**, in which the ion beam is focused to a range of only a few nanometers by means of electrostatic lenses and apertures. The secondary electrons produced as a result of the interaction of the ion beam with the sample also allow the etching process to be imaged and checked as in an SEM. This allows the etching process to be continuously imaged and followed with very high accuracy, which is particularly crucial for the target preparation of very small details, such as defect structures in nano-electronics transistors. The etching process is carried out with gallium or in plasma-based beam sources with the aid of the noble gas xenon, whereby the etching rate increases by a factor of 10–20 due to the higher mass of the ions. For the TEM cross-section preparation, trenches are etched out directly in front of and behind the sample region of interest. The resulting sample lamella is thinned in subsequent polishing processes using the ion beam to thicknesses below 200 nm to 10 nm to achieve the electron beam transparency required for TEM. Once the target thickness is reached, the ion beam completely separates the sample lamella from the sample at its side and bottom edges – see Fig. 3.3. With the aid of a needle manipulator or clip gripper, the lamella is picked up, guided out and placed on a TEM sample holder.

3.3   Preparation of Soft Matter (By Cryo-Ultramicrotomy)

In the ultramicrotome, a mechanical separation technique is used to produce thin and ultra-thin sections with a glass or diamond knife. The specimen holder of the ultramicrotome moves along a sharp knife edge (glass or diamond knife) and produces a section with a thickness determined by the forward movement of the specimen holder (see Fig. 3.4). The forward movement of the specimen arm can be by a thermal or mechanical feed. The cuts are either collected directly on the knife surface ("dry cutting") or collected in a water trough ("wet cutting"). Cutting can also be performed at lower temperatures down to the temperature of liquid nitrogen (cryo-ultramicrotomy). In cryo-cutting, softer materials are cooled to below their glass temperature and thus hardened (fixed). Softer materials can also be hardened chemically or physically by irradiation, by initiating cross-links in the material (often such hardening reactions also produce selective contrast enhancements – see below). The principle of ultramicrotomy has remained the same over the years, but user-friendliness and operability have been significantly improved.

New knife configurations, such as the so-called "oscillating knife" from Diatome Ltd., Biel (Switzerland), which is oscillated by a piezo crystal, have also extended the range of applications of microtomy [23]. The oscillating knife reduces smearing effects and compression or distortion of the structure when cutting ductile materials (see Fig. 3.5) [24]. The oscillating knife can also lead to significant improvements in cryo-ultramicrotomy.

3D analyses (spatial analyses) of structures can be performed by serial sections with the ultramicrotome and subsequent examination of the individual sections in

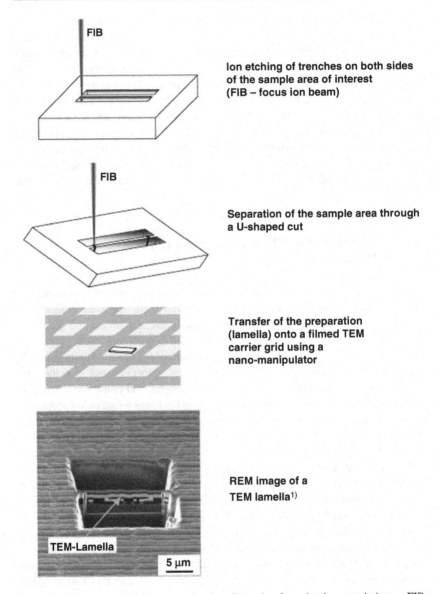

Ion etching of trenches on both sides of the sample area of interest (FIB – focus ion beam)

Separation of the sample area through a U-shaped cut

Transfer of the preparation (lamella) onto a filmed TEM carrier grid using a nano-manipulator

REM image of a TEM lamella[1]

Fig. 3.3 Direct production of transparent lamellae using focus ion beam technique – FIB. (From [9], Source: Fraunhofer IMWS Halle)

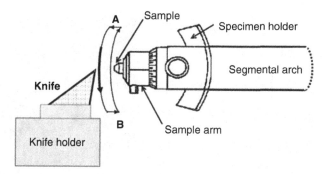

Fig. 3.4 Schematic representation of the relative movement of specimen and knife during the cutting process in an ultramicrotome (A – feed, B – retraction of the specimen arm). (From [8])

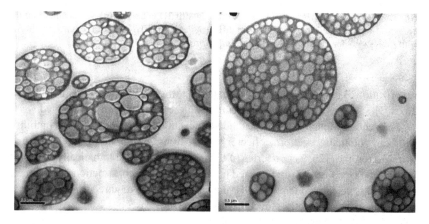

Fig. 3.5 TEM images of impact-resistant polystyrene with so-called salami particles (selectively stained) in a PS matrix: deformed and compressed on the left after cutting in an ultramicrotome with a conventional knife, on the right after cutting with an oscillating knife. (With kind permission of C. Mayrhofer, Graz)

the TEM in conjunction with image processing. Two newly developed methods have automated this process and thus greatly simplified it, although images can only be obtained in the SEM with the lower resolution compared to the TEM. In the first method, an ultramicrotome is built into the sample chamber of an SEM (see Fig. 3.6) [25]. The ultramicrotome is used to cut sections and the freshly created cutting surface is imaged with the SEM or with an ESEM to avoid conductive

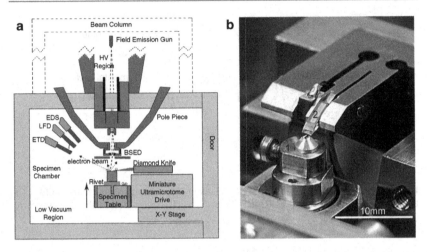

Fig. 3.6 Ultramicrotome in an SEM for direct in-situ 3D analyses: (**a**) Schematic of an ultramicrotome mounted in an ESEM with different detectors, (**b**) View of the ultramicrotome with sample holder (1) and diamond knife (2). (From [25])

vaporization. The process is then repeated automatically up to several thousand times. The images of the successive cutting surfaces are spatially assembled.

The second method uses an extended conventional ultramicrotome combined with an automatic section collector (ATUMtome, Automatic Tape-collecting UltraMicrotome) [26, 27]. A series of ultrathin sections is automatically collected from a tape (see Fig. 3.7) and imaged in the SEM. The reconstruction of the image data in 3D is performed in powerful computers. The time required for image acquisition can be significantly reduced with special microscopes, such as the multibeam SEM (ZEISS MultiSEM – see Chap. 5).

Serial cutting is also possible with an FIB, but removing material with an FIB takes much more time than for an ultrathin cut with an ultramicrotome.

3.4   Contrast Enhancements

More often, the contrast between the structural details of interest is too low during observations in the TEM or SEM, or should be specifically improved. Several methods are available for this purpose, which are based on either chemical or physical effects [10].

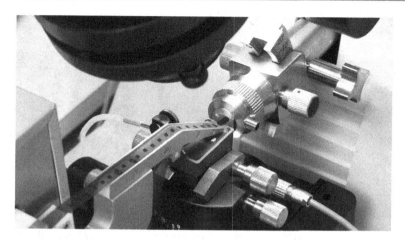

Fig. 3.7 ATUMtome: an ultramicrotome with an automatic section collector (long band in the foreground; diamond knife in the middle, sample with movable arm in the upper right). (Source Science Services GmbH Munich; RMC Boeckeler Tucson)

Chemical treatment (staining) can be carried out with various reagents which cause cross-linking or selective addition of heavier elements and thus a local increase in density. Common chemical staining (and thus mostly fixing) media are osmium and ruthenium tetroxide, uranyl acetate, bromine and combinations with other chemicals. Typically, these attack double bonds of the material, leading to cross-linking or accumulation of heavier elements. The broadest applications of such chemicals are in biology, medicine, and polymer investigation. An example from biology is shown in Fig. 3.8 with bacteria after different fixation and staining steps. For materials with different components, combined staining with successively different reagents is sometimes successful (cf. Fig. 4.1). Chemical treatment can be carried out before preparation (e.g. cutting in the ultramicrotome) or after preparation on the thin sections.

Two **physical processes** can also lead to contrast enhancement. In heterogeneous and easily crosslinkable materials such as polymers, various primary and secondary effects can be initiated by gamma rays or electron beams, which lead to selective contrast enhancement via mass loss or crosslinking. Figure 3.9 shows the effect of contrast enhancement by γ-irradiation in a semi-crystalline polymer (LDPE, low density PE) by the emergence of concentric rings in the spherulites (Fig. 3.9a) and increased visibility of crystalline lamellae (bright in Fig. 3.9b). This

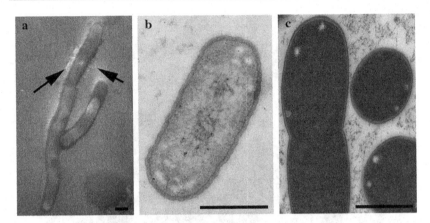

Fig. 3.8 Bacteria in EM (μm line corresponds to about 0.5 μm each): (**a**) untreated in ESEM (Shewanella putrefaciens), (**b**) chemically fixed and stained, TEM (Xanthomonas campestris), (**c**) after cryofixation/cryosubstitution, stained, TEM (Escherichia coli). (Courtesy of G. Hause [29])

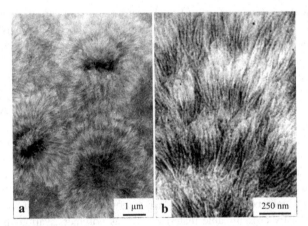

Fig. 3.9 Contrast development in a semi-crystalline polymer (LDPE) by γ-irradiation (with a dose of 20 MGy): (**a**) concentric rings within the spherulites, (**b**) radially arranged lamellae within the concentric rings. (From [21])

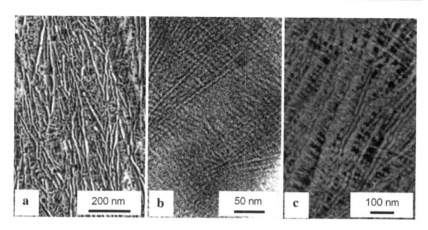

Fig. 3.10 Comparison of the results of different preparation and examination methods on a plastic (α-iPP), (**a**) chemically etched surface (permanganate etching) in SEM, (**b**) chemically stained ultrathin section in TEM, (**c**) surface without pretreatment, AFM image. (From [30])

effect has been termed **"irradiation-induced contrast enhancement"** (cf. also Fig. 2.12) [21].

An unconventional method revealed by in-situ microscopy is the stretching of thin heterogeneously structured materials. Softer or more easily deformable areas (smaller modulus of elasticity) are stretched more under load and appear brighter in the TEM image. This effect, known as **"strain-induced contrast enhancement"** [21], is also valuable when chemical contrasting fails (e.g. in chemically resistant plastics).

Sometimes several different methods and procedures can be used in structural investigations, which of course give the same results but with different details. A comparison of the results of different electron microscopy techniques and preparations on one plastic is shown in Fig. 3.10. The typical "cross-hatched" morphology of isotactic polypropylene (iPP) is reproduced in all cases, although the detail detectability of the lamellae differs.

Image Processing and Image Simulation

4

Methods of image processing, image reconstruction and image analysis aim to modify the image information of a microscopic image in such a way that both the image quality is improved and quantitative information on structural details can be determined [31]. The prerequisite is a digitized image, which today is directly obtained and stored in electron microscopes and scanning probe microscopes.

The term **image processing** subsumes all techniques that directly modify image information in a digital or analog manner without modeling the imaging process itself. The simplest image processing is image improvement, where image details are made more discernible and simple imaging errors are corrected, such as manipulations of contrast and image brightness. Image restoration is a bit more sophisticated and involves eliminating or reducing image degradation, blurring, and washes. Noise reduction is achieved by image accumulation, i.e. by combining several well-adjusted images.

An **image analysis** is used to quantitatively determine certain image information of interest, such as phase fractions, particle sizes or particle distances. In the case of clear grey value separations, automatic image analysis is usually possible. Figure 4.1 shows a TEM image of a three-phase polymer blend, where the corresponding phases are reflected by defining the grey value regions and the phase fractions are determined.

A central issue of all microscopic examinations is to obtain spatial structural information from the two-dimensional images. The classical technique is to tilt the specimen in the microscope by a defined angle and then view the two stereo images – in the same way as the two eyes create a spatial impression by viewing the object at slightly different angles (see Fig. 4.2).

G. H. Michler, *Compact Introduction to Electron Microscopy*, essentials, https://doi.org/10.1007/978-3-658-37364-1_4

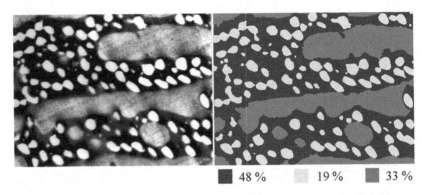

| ■ 48 % | ■ 19 % | ■ 33 % |

Fig. 4.1 TEM image of a polyamide blend in which the three phase fractions (PA 6, PA 12, modifier) can be identified in different grey values by selective chemical staining (left) and the phase fractions were determined by assigning defined grey values to the phases (right). (From [9])

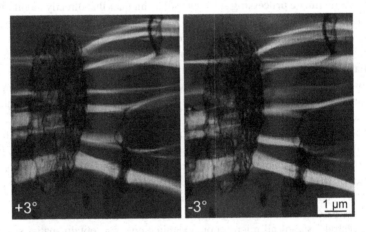

Fig. 4.2 Semi-thin sections of a stretched polymer sample (HIPS: rubber particles stained with OsO_4 in a PS matrix) at tilt angles of $+3°$ and $-3°$: different widths of the crazes (light bands) show their thickness and height. (Images taken in 1000 kV HEM with a strain tilt holder, strain direction vertical, tilt axis horizontal). (From [30])

Special tilting stages allow the specimen to be tilted in the microscope by more than $\pm45°$ and thus allow serial images of increasingly tilted specimens to be taken [22, 30]. The merging of the individual images to form the overall impression of a tilted specimen is carried out using 3D methods. Further possibilities are the production of numerous successive sections or thin sections with an ultramicrotome, the subsequent automatic image acquisition in the SEM, the computer-assisted evaluation of the individual images and the composition to form a 3D representation (cf. Fig. 3.6 – with the automatic production and acquisition of sections with a

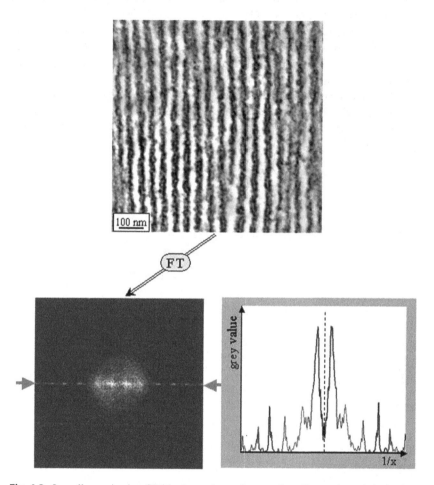

Fig. 4.3 Lamella spacing in a SB block copolymer (styrene-butadiene polymer) determined by image processing. (**a**) TEM image of a stained ultrathin section – PS lamellae light, PB lamellae dark stained, (**b**) Fourier transformation of the image, (**c**) grey value profile of the line marked in (**b**) provides the averaged lamella spacing of the original image. (From [9])

UM in the SEM – and Fig. 3.7 – by means of an ultramicrotome section collector and analysis with a multibeam SEM (see Sect. 2.3)).

Important techniques of image processing are based on optical diffraction and filtering of images. A Fourier transform of the digitized image calculated in the computer corresponds to an optical Fraunhofer diffraction. This allows symmetries of the image to be read directly or structures present in the image to be highlighted by means of suitable filters (see Fig. 4.3). The Fourier transformation of the image also allows imaging errors of the electron microscope to be made directly visible or preparation artefacts (scratches, cutting defects from ultramicrotomy) to be removed. The reverse transformation then provides a "corrected" image in which the corresponding structures are clearly visible.

In addition, image processing and analysis are used to determine starting models for image simulation. **Image simulation is** used to calculate the image formation in the electron microscope. This is particularly necessary for the interpretation of high-resolution images.

Outlook

5

The question of further developments and challenges in electron microscopy arises essentially from the requirements in the life and material sciences. In the material sciences, property improvements are increasingly sought through optimizations on a nanoscopic and atomic level. On the other hand, a stronger coupling of micro-structural information with chemical and other analytical results is aimed at. Besides the improved analysis of the microstructure down to the atomic size level, diffraction analysis, spectroscopy and the imaging of internal fields, the observation of dynamic processes and phenomena in physics and medicine is gaining further importance. This extends, for example, to ultrafast electron microscopy with femtosecond resolution. Furthermore, the sample throughput in preparation and in the electron microscope must be improved by automating routine steps. For example, brain research is interested in improved and faster electron micrographs of brain tissue at different stages of disease. Changes in tissues and interactions of bacteria with organs should be analyzed under conditions as close to life as possible. From these and other examples also mentioned in the preface, demands on electron microscopy are derived.

As an example, the combination of a preparation method for rapid specimen preparation (ultramicrotomy with an automated section collector – see Fig. 3.7) with a faster imaging technique (multibeam SEM – see Sect. 2.3) is shown in image montage Fig. 5.1 from an experiment for brain research. On the left is a light microscopic overview image of the entire sample holder, on which the individual brain slices were computer-assisted detected and marked. On the right, a data set of

G. H. Michler, *Compact Introduction to Electron Microscopy*, essentials, https://doi.org/10.1007/978-3-658-37364-1_5

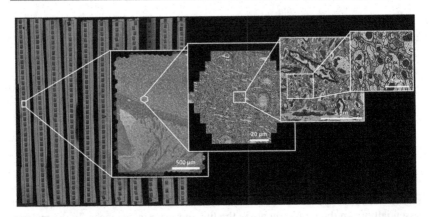

Fig. 5.1 Example of increasing sample throughput (rapid production of thin sections and microscopic images) from an experiment for brain research. The thin sections (left) were produced with an ultramicrotome with automated section collector, the SEM images were obtained with a multi-beam SEM (ZEISS MultiSEM). (The sample was kindly provided by Jeff W. Lichtman, Harvard University, compilation of images by A. Eberle, Carl Zeiss Microscopy GmbH, Germany). (From [13, 14, 27, 28])

a complete mouse brain section created with the ZEISS MultiSEM is shown in gradually increasing magnification.

In addition to technical developments, however, the training of microscopy personnel is essential and should not be neglected in order to be able to make full use of the increased performance of microscopes.

What You Learned from This *essential*

- What electron microscopy is and what is the advantage of electron microscopy compared to classical light microscopy.
- What distinguishes electron beams from light beams and why atoms can be made visible.
- What are the characteristics of transmission electron microscopy and scanning electron microscopy. What electron microscopes look like and how they work.
- Which materials can be examined with which microscopy technique and how they have to be prepared for the examination.
- Numerous findings in the materials and life sciences are based on the use of electron microscopy.
- What future demands on techniques and procedures can be expected from electron microscopy.

G. H. Michler, *Compact Introduction to Electron Microscopy*, essentials, https://doi.org/10.1007/978-3-658-37364-1

References

1. Gloede, W. (1986). *Vom Lesestein zum Elektronenmikroskop.* VEB Verlag Technik.
2. Knoll, M. (1935). *Zeitschrift Technik Physik, 11,* 467.
3. v. Ardenne, M. (1938). *Zeitschrift für Physik, 109,* 553.
4. v. Ardenne, M. (1938). *Zeitschrift Technik Physik, 19,* 407.
5. v. Ardenne, M. (1940). *Elektronen-Übermikroskopie.* Springer.
6. Henneberg, W., & Recknagel, A. (1935). *Zeitschrift Technik Physik, 16,* 621.
7. Bethge, H., & Klaua, M. (1983). *Ultramicroscopy, 11,* 207.
8. Müller, E. W. (1951). *Zeitschrift für Physik, 131,* 136.
9. Michler, G. H., & Lebek, W. (2004). *Ultramikrotomie in der Materialforschung.* Hanser.
10. Michler, G. H. (2008). *Electron microscopy of polymers.* Springer.
11. Michler, G. H., & Katzer, D. (Hrsg.). (2017). Elektronenmikroskopie in Halle (Saale) – Stand, Perspektiven, Anwendungen. Bethge-Stiftung Halle.
12. Lichte, H., & Lehmann, M. (2008). Electron holography – Basics and applications. *Reports on Progress in Physics, 71,* 016102.
13. Eberle, A., et al. (2015). *Journal of Microscopy, 259,* 114.
14. Carl Zeiss Microscopy GmbH, Germany. www.zeiss.com/multisem
15. Telieps, W., & Bauer, E. (1985). *Ultramicroscopy, 17,* 57.
16. Höfer, A., Duncker, K., Kiel, M., & Widdra, W. (2011). *IBM Journal of Research and Development, 55,* 4.
17. SPECS Surface Nano Analysis GmbH, Berlin. www.specs.com
18. Binnig, G., Rohrer, H., Gerber, C., & Weibel, E. (1982). *Physical Review Letters, 49,* 57.
19. Bethge, H., & Heydenreich, J. (Eds.). (1982). *Elektronenmikrokopie in der Festkörperphysik.* VEB Deutscher Verlag der Wissenschaften.
20. Bethge, H., & Heydenreich, J. (Eds.). (1987). *Electron microscopy in solid state physics.* Elsevier.
21. Michler, G. H. (1992). *Kunststoff-Mikromechanik: Morphologie, Deformations- und Bruchmechanismen.* Hanser.

22. Messerschmidt, U., Appel, F., Heydenreich, J., & Schmidt, V. (1990). *Electron micros-copy in plasticity and fracture research of materials*. Akademie.
23. Studer, D., & Gnaegi, H. (2000). *Journal of Microscopy, 197,* 94–100.
24. Mayrhofer, C. Center of electron microscopy, Graz Austria.
25. Zankel, A., Kraus, B., Poelt, P., Schaffer, M., & Ingolic, E. (2009). *Journal of Micros-copy, 233,* 140–148.
26. Webster, P., Bentley, D., & Kearney, J. (2015). *Microscopy and Microanalysis, 17,* 17–20.
27. Science Services GmbH, München.
28. RMC Boeckeler.
29. Hause, G., & Jahn, S. (2010). Molecular and cell biology methods for fungi. *Methods in Molecular Biology, 638,* 291–301.
30. Michler, G. H. (2016). *Atlas of polymer structures: Morphology, deformation and frac-ture structures*. Hanser.
31. Heydenreich, J., & Neumann, W. (Eds.). (1992). *Image interpretation and image pro-cessing in electron microscopy*. Halle.

Further Reading[1]

32. Williams, D. B., & Carter, C. B. (1996). *Transmission electron microscopy: A textbook for materials science*. Plenum.
33. Goodhew, P. J., Humphreys, F. J., & Beanland, R. (2000). *Electron microscopy and anal-ysis* (3rd ed.). Taylor & Francis.
34. Fultz, B., & Howe, J. (2003). *Transmission electron microscopy and diffractometry of material of material* (2nd ed.). Springer.
35. Staniforth, M., et al. (2002). *Scanning electron microscopy and X-ray microanalysis*. Kluwer Academic & Plenum.
36. Ernst, F., & Rühle, M. (Eds.). (2003). *High-resolution imaging and spectrometry of ma-terials*. Springer.
37. Umbaugh, S. E. (2005). *Computer imaging: Digital image analysis and processing*. Tay-lor & Francis.
38. Schönherr, H., & Vancso, G. J. (2010). *Scanning force microscopy of polymers*. Springer.
39. Carter, B., & Williams, D. (Eds.). (2016). *Transmission electron microscopy: Diffrac-tion, imaging, and spectroscopy*. Springer.
40. Goldstein, J. I., Newbury, D. E., Michael, J. R., Ritchie, N. W. M., Scott, J. H. J., & Joy, D. C. (2018). *Scanning electron microscopy and X-ray microanalysis*. Springer.

---

[1] There is a very extensive literature on electron microscopy, which has been updated again and again over many years, and which either outlines the field in general or concentrates on special techniques. The following monographs should be mentioned as standard works (cf. also [9, 10, 19, 20]):

Printed in the United States
by Baker & Taylor Publisher Services